Yusop Masdal

A survey of current socio-economic condition of muslims in Zamboanga City and Basilan

GRIN Publishing

Bibliographic information published by the German National Library:

The German National Library lists this publication in the National Bibliography; detailed bibliographic data are available on the Internet at http://dnb.dnb.de .

Imprint:

Copyright © 2009 GRIN Verlag, Open Publishing GmbH
Print and binding: Books on Demand GmbH, Norderstedt Germany
ISBN: 978-3-640-77404-3

This book at GRIN:

http://www.grin.com/en/e-book/162381/a-survey-of-current-socio-economic-condi-tion-of-muslims-in-zamboanga-city

A SURVEY OF CURRENT SOCIO-ECONOMIC & POLITICAL ROUTINE OF MUSLIM COMMUNITIES IN ZAMBOANGA CITY AND BASILAN

A Research Study by Yusop B. Masdal

I. Introduction

This study is aimed at thoroughly examining everyday socio-economic and political routines of Muslims in Zamboanga City and Basilan, with emphasis on economic issues, in order that a wider and more conclusive perspective may be gained on these aspects, obtaining direct observations on the issues presented herein.

As a beneficial consequence, this study would bring forth highly-verified information that will be entirely useful and significant to the development efforts in Muslim Mindanao and therefore contributing in a palpable manner to the peace efforts thereat.

When the desired information and data would be collated in the final results of this study, government and other concerned entities --- such as developmental non-governmental organizations and international funding agencies --- could gain highly provable insights of routines and daily socio-economic and political activities or routines (especially on economic or livelihood activities) of Muslims and members of ethnic minorities in Zamboanga City and Basilan, thus, allowing a more effective creation of economic opportunities for the subjects of this study, gaining initial foothold on a series of economic activities that the respondents are more accustomed to, and are most convenient with.

II. Background of the Study

It is well-known verity that continuous aids and sizable funding have been injected to the peace-building efforts in Mindanao, a region that was torn and still continues to suffer decades of war and strife between secessionist groups aiming at independent territory and government military forces.

It is also of wider nuance that often, these livelihood programs often goes to naught or had resulted to low degree of success as their social impact were limited to a minimum.

It is the premise that this study aims to take-off, where livelihood programs and other economic developmental strategies for Muslim Mindanao should be designed and programmed in accordance with available resources at hand and more importantly should be attuned to the routine and daily economic activities of respondents, allowing easier integration of process and methods, like being fish in a water, where economic assistance would mean helping the subjects embark on economic endeavors that they are more keen or adept at, and not artificially locate them into alien activities that they are not familiar with in the first place, having no prior experience or being alien to such economic endeavors.

The main aims of this study would be:

1. *The determination of the best and most effective manner of giving economic assistance to Muslims in Zamboanga City and Basilan.*

2. *How to help and assist them on economic opportunities that they are more familiar with, so that there would be proper matching of aids and capability of respondents, not forcing them or inducing them to have access on funding only to fail in*

Thus, routine social aspects of the respondents would subsequently be collated and defined as well as their political views and insights, for economic activities are directly related to the social surroundings and political environment.

It is of great note that the lack of livelihood activities or opportunities among Muslims and members of ethnic minorities in Mindanao remains to be one of the biggest factors in the negation of a lasting peace solution in Mindanao.

An effective economic strategy in this region should be one of the best options in resolving long-lasting conflicts and issues there.

III. Conceptual Framework

Figure 1.1 exhibits the basic structure of this study, on a step-by-step and chronological manner. First to be identified are aims and goals, such end-results as poverty-reduction among the population queried upon in this study, as one among several means of reducing instances for conflict in Mindanao, as war and chaos are on huge part due to economic discontentment.

The means of obtaining the end-results are mostly on enhancement of economic capacity. To gain high overview on this, the study seek first to identify the most prevalent livelihood activity of the respondents and then upon that premise, the study thereon aims to generate information on how to most effectively alleviate economic conditions of the respondents based on data and information gathered about their basic and general socio-economic and political routines.

And lastly and most importantly, the study aims to find means and methods that would lead to the attainment of the aims and goals of the study.

FIGURE 1.1

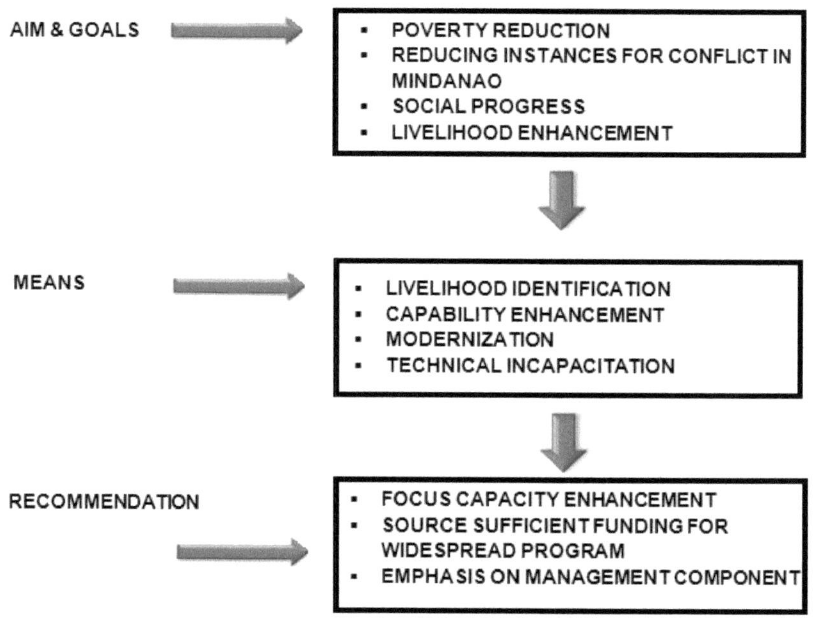

IV. Review of Related Literature

The review of related materials for this research work will focus on extant economic theories that the researcher deems to be critical in the resolution of the hypotheses involved. Also included are economic policies that have been applied already, experiences that are related to the main issues of this study and previous studies made with aims and contents similar to this work.

A.) ECONOMIC THEORIES and PRINCIPLES

a.)   **COMPARATIVE ADVANTAGE**[1] - is a principle in international trade that expresses the fact that the efficiency of country A relative to country B, varies from product to product.

A country has comparative advantage in products for which its relative efficiency is highest.

Comparative Advantage as to Factor Endowment states that a country tends to have a comparative advantage in products using factors which are relatively abundant and cheap in that country.

In simple terms, a country must produce primarily products or commodities that it has comparative advantage over other countries and not those that it lacks of. The originator of this trade theory, David Ricardo[2], had defined and explained Comparative Advantage in an example involving England and Portugal. In

---

[1] Lloyd G. Reynolds, "ECONOMICS : A General Introduction", Yale University, HOMEWOOD Publishing, Illinois 60430

[2] David Ricardo, "On the Principles of Political Economy and Taxation", London: John Murray, 1821. First published 1819.

Portugal it is possible to produce both wine and cloth with less labor than it would take to produce the same quantities in England. However the relative costs of producing those two goods are different in the two countries. In England it is very hard to produce wine, and only moderately difficult to produce cloth. In Portugal both are easy to produce. Therefore while it is cheaper to produce cloth in Portugal than England, it is cheaper still for Portugal to produce excess wine, and trade that for English cloth. Conversely England benefits from this trade because its cost for producing cloth has not changed but it can now get wine at a lower price, closer to the cost of cloth. The conclusion drawn is that each country can gain by specializing in the good where it has comparative advantage, and trading that good for the other.[3]

b.) **REVEALED COMPARATIVE ADVANTAGE**[4] - we can learn more about comparative advantage by examining what countries actually export and import. Professor Bela Balassa has made such an analysis for the major industrial countries:

Consider this formula:

$$\frac{\textbf{COUNTRY A EXPORTS of PRODUCT A / WORLD EXPORT OF PRODUCT A}}{\textbf{COUNTRY A EXPORTS OF ALL MANUFACTURE / WORLD EXPORT OF ALL MANUFACTURE}}$$

The result would be the "revealed comparative advantage index" of country concerned.

---

[3] WIKIPEDIA : http://en.wikipedia.org/wiki/Comparative_advantage

[4] Bala Balassa and Associates, Studies in Trade Liberalization (Baltimore, John Hopkins Press, 1967) Chapter 1.

Using this method, TABLE 1.1 exhibits the top three countries in terms of revealed comparative advantage and the leading products of each.

Rank	UNITED STATES	UNITED KINGDOM	JAPAN
1	Aircraft	Tractors	Footwear
2	Agricultural machinery	Power-generating machinery	Cotton yarn
3	Office machinery	Paints/ varnishes	Synthetic fabrics
4	Power-generating machinery	Explosives	Pottery
5	Photographic and cinematographic materials	Buses, trucks	Cotton fabrics
6	Inorganic chemicals	Woolen fabrics	Buses, trucks
7	Railway vehicles	Bicycles	Woolen fabrics
8	Electrical power equipment	Nickel, wrought	Ships. Boats
9	Bodies, chassis and frames	Copper, wrought	Musical instruments
10	Scientific. Medical and optical	Fur clothing	Travel goods, handbags

c.) **PORTER'S DIAMOND** – In his work "The Competitive Advantage of Nations"[5], Michael Porter explained and discusses on how a nation can viably be competitive in the world market, by learning to upgrade and update and keeping up with the rest of the competitors.

He had emphasized that industries prosper due mainly to a vibrant home environment, one that is forward looking, dynamic and challenging. Porter enumerates four attributes that would enhance a nation's competitiveness, what is termed as the 'diamond of national advantage', namely factor conditions,

---

[5] Porter, M. E. The Competitive Advantage of Nations. New York: Free Press, 1990. (Republished with a new introduction, 1998.)

demand conditions, related and supporting industries and firm strategy, structure, and rivalry.

These four attributes should interplay to hone a nation's trade and economic advantage.

B.) DEVELOPMENTAL STUDIES, POLICIES AND APPROACHES

a.) **REASONS FOR EAST ASIAN ECONOMIC PERFORMANCE**[6] - in analyzing economic performance especially for Taiwan and Korea, analysis of developmental strategies, of forecasts and intricacies and assumptions about probable causal sequences held by policy makers when they make economic decisions. Such analysis involved:

> 1)   *Estimating the production function of Third World countries by identifying the source of output growth like capital, labor and technical progress.*
>
> 2)   *Identifying the factors responsible for these sources of growth.*
>
> 3)   *Identifying the economic policies needed to bring about these factors and their efficient use.*

b.) **STRATEGIES FOR THE DEVELOPMENT and MANAGEMENT of AGRICULTURE**[7] - studies in Third World agricultural economies showed

---

[6] Chapter 2, "Comparative Political Economy of East & South Asian : A Critique of Development Policy and Management" by R.C. Mascarenhas 1999, Great Britain, Macmillan Press LTD

[7] Chapter 10, "Comparative Political Economy of East & South Asian : A Critique of Development Policy and Management" by R.C. Mascarenhas 1999, Great Britain, Macmillan Press LTD

successful agricultural modernization process and state intervention. Two important characteristics of East Asian agriculture are the impact of Japanese modernization under colonial rule and the subsequent state-directed strategy based on a successful program of land reform.

The experience of South Asia was dissimilar were agriculture stagnated and lapse under colonial negligence and inattention.

c.) **MODELS OF AGRICULTURAL DEVELOPMENT**[8] - In a study titled "Agricultural Development : An international perspective" Vernon Wesley Ruttan and Yujiro Hayami enumerates and distinguishes five (5) models of agricultural development namely Resource Exploitation Model, Conservation Model, Urban Industrial Impact Model, the Diffusion Model and the High-Payoff Input Model.

Collectively known as "induced Innovation and Agricultural Development", Ruttan and Hayumi have this outline of their hypothesis: [9]

"The [induced innovation] model attempts to make more explicit the process by which technical and instititutional changes are induced through the responses of farmers, agribusiness entrepreneurs, scientists, and public administrators to resource endowments and to changes in the supply and demand of factors and products. The state of relative endowments and accumulation of the two primary resources, land and labor, is a critical element in determining a viable patterns of technical change in agriculture. Agriculture is characterized by much stronger constraints of land on production than most other sectors of the economy.

[8] Economic and Social Development : Trends, Problems, Policies by Adam Szirmai, Prentice Hall, Europe 1997

[9] WIKIPEDIA : http://en.wikipedia.org/wiki/Vernon_Wesley_Ruttan

Agricultural growth may be viewed as a process of easing the constraints on production imposed by inelastic supplies of land and labor. Depending on the relative scarcity of land and labor, technical change embodied in new and more productive inputs may be induced primarily either (a) to save labor or (b) to save land."

C.) CHINA'S ECONOMIC DEVELOPMENT STRATEGIES[10] - China is now an economic giant and its economy is expected to further expand and develop in the coming years. In the nascent stage of this economic phenomenon was the agricultural reconstruction in the mid-1980's utilizing the following approaches:

1)   *Suburban Agriculture – Local and provincial governments there had greatly encouraged residents of suburban areas to grow agricultural products to feed the urban market. The suburban market in Shenyang was a successful model of suburban farming. Many suburban farmers now supply agricultural produce and sideline goods which in turn give these farmers an opportunity to increase their standard of living.*

2)   *Price Reform – taking away any form of price control of agricultural produce by the government.*

3)   *Modernizing Farming Practices – recognizing agriculture's significance to the national health and political stability issues, the government in China continues to subsidize rural people. Introduction of modern technologies yields higher income for farmers and resulted to improved standard of living.*

---

[10] CHINA'S ECONOMIC DEVELOPMENT STRATEGIES FOR THE 21st CENTURY by Harry J. Waters – 1997 Quorom Books, 88 Post Road West, Westport, CT 06881

4) *Continuing Education for Farmers – In the aim of maximizing gains, farmers are given continuing education of techniques and management, opening robust career in farming. The training, the so-called Green Certificate training had been carried out in 500 counties over there.*

5) *Economic Alternative to Farming – aiming to change the rural landscape of its central and western areas, China government had embarked on an industrialization program that provided alternative economic livelihood to farmers. Inviting foreign investors to locate in trade zones had dramatically boosted this program. Another approach was encouraging every provincial government involved to readjusting its industrial setup, creating industrial districts within their area.*

D.) SOUTH KOREA'S AGRICULTURAL SUBSIDY[11] – South Korea's subsidy program of its farmers is one prime example in such economic method, which had resulted to high-productivity for farmers and thereon very high food sufficiency for the country, redounding to political stability that had allowed authoritarian administration to wrestle South Korea's economy from 'poor agrarian' in the 1960's to 'highly-industrialized' today.

---

[11] http://findarticles.com/p/articles/mi_m3809/is_n63/ai_11174350/

V. Methodology

A. Focus Group Discussions

This research work utilizes Focus Group Discussions (FGDs) as the main tool for information gathering.

In these group conversations, issues pertinent to the objectives and aims of the study are well-ventilated and deeply scrutinized. In this method, a forum is thereby formed allowing free flowing exchange of ideas with unhindered interpellations and cross-questioning.

The researcher seeks mainly to collect raw information without possibly the intervention of sophisticated and educated views, enjoining primarily participants that are target respondents of this study, which are the ordinary and routine members of Muslim communities within the area of study, neutral of gender and within the range of ages where individuals are mostly expected to make a living, approximately from age 18 to 60.

The Focus Group Discussions were convened for actual conversation, maintaining free flowing of ideas and sentiments. The researcher presented four (4) theme questions, to wit;

1. What are your livelihood activities and what are the most prevalent livelihood activities in your area?

2. What kind or form of government assistance that best suits your needs, in relation to your livelihood activities?

3. What is your view or opinion to the current Mindanao
    peace process?

4. Give statements about everyday routine in your
    community.

Probing questions were supplemented to support the main responses such as for example the fishing methods or tools that are usually utilized for fisherfolks, the cause or causes on out-of-school youth instances, the amount of time for agar-agar to be harvest, the occurrence of privately-owned vehicles in the area, the ages when individuals start to make a living, etc..

Actual Steps/Means in Conducting FGD:

1) The researcher proceeds to target location and ask permission from local LGUs.

2) Designation of point man. Point man is oriented as to targeted respondents or participants.

3) The researcher opts for ordinary individuals without any prior guided information about the issues to be discussed, and limiting barangay officials to minimum.

4) Scheduling of meeting – time and venue.

5) Each participant is given fifty pesos each, minimum of 10 participants and maximum of fifteen. Foods and drinks are part of the cost. Point man is paid higher fee, two hundred pesos.

6) The discussion is held in public areas like eating places or barangay halls.

7) Conversations are recorded thru voice recorder.

8) After the discussions were held, researcher makes follow up visits to clarify or confirm matters.

9) A verbatim account of each FGD is done.

10) Synthesis and summaries are subsequently done.

11) Overview and conclusion.

Focus Group Discussions were held for the following areas:

1. MARIKI – a Muslim-dominated community located within 3 kilometer from City Hall, south of the city. Houses on stilts or built above seawater are easily observed and it has very high percentage of Muslim residents, amounting to almost 97% of its residents. Mariki began in 1975 as Sahaya Village program by the government, a socio-economic component of the government's anti-insurgency campaign then against the rising and consolidating strength of the Moro National Liberation front (MNLF). Most of the residents there were migrants from the chaotic neighboring islands and provinces during those turbulent years.

Estimated population is 5,334 (2007 estimate)[12]

2. RECODO – A seaside baranggay located northwest of the City hall and about 10 kilometers in distance, with a population of 17,754 consisting mostly Muslims with about 30% Christians, especially in areas near the sea coast. It is an area in the city where ships are built and repaired as its coast are always filled with ships in dock. Sardine factories are also located there that most residents make their living supplying fish to them and/or working as sardine factory workers.

---

[12] City of Zamboanga LGU official website : URL http://www.zamboanga.gov.ph/

Most of the residents here particularly those living near the shoreline, are fisherman by occupation.

3. TALUKSANGAY – A traditional samal-bangingi tribal hometown, even from the beginning of modern history, Taluksangay is situated in northeast portion of the city and about 19 kilometers from the city hall. Due to its population make-up, nearly 100% Muslims, it is said to be one of a few barangays that does not celebrate fiesta. According to the website of Zamboanga City local government, it is where Islamic culture was first introduced in 1885 and where the first mosque of the city was constructed there. Total population is at 7,116 with 1,266 households.

4. BASILAN – 80% of the nearly half-million population of Basilan is composed of Muslims. It is an island province that lies south of Zamboanga City and this proximity allows brisk trading and people movement between Zamboanga and Basilan. Except for its capital town Isabela, Basilan is mostly part of the Autonomous Region in Muslim Mindanao. The southern areas like Sumisip and Tuburan - where mostl Muslims resides --- are much-less developed in economic terms and are prone to armed-conflict.

B. Interviews

One-on-one and group interviews are also applied as approaches in data gathering, augmenting significantly the information initially collated thru Focus Group Discussions.

In this method, the researcher visits Muslim communities and randomly interviews individuals who are residents thereof, neutral of gender while targeting age range where individuals are already expected to enter livelihood activities.

The interviews are done on open location that not only the interviewees are present but other passersby or nearby people, who often affirms or intervene in conversations. Officials of LGU's within the community as well as traditional leaders are also interviewed to support and verify data gathered thru FGDs.

Areas where interviews were conducted:

1. Sacol Island – an island with four barangays within Zamboanga City jurisdiction, namely Busay, Pasilmanta, Landang Laum and Landang Gua. Farming is prevalent there just as fishery is.

2. Sirimon islands – is part of the 11 Islands group off the eastern coast of Zamboanga City. It was known as "No man's land" during the 1970's when the government's war with the MNLF was at its peak. Most residents live off the sea as a mode of livelihood. The Act For Peace Programme (ACT MINDANAO) of the Mindanao Economic Development Council (MEDCo)[13] has just completed aid program there were 200 houses were built for the residents and 30 pump boats were given as aid for fisherfolks.

3. Panubigan – 79% of the residents are Muslims with a thriving seaweeds farming activity, along with coconut plantations.

4. Campo Islam – a Muslim-dominated barangay just about 3 kilometers north of City Hall. Comprising of 75% Muslims, the community started as a reservation area

---

[13] ACT FOR PEACE PROGRAMME URL : http://www.actforpeace.ph/component/option,com_frontpage/Itemid,1/

declared during the American period, due especially to Muslim residents. Total population is 11,523 with 1,580 households.

5. Arena Blanco – 97% of the residents comprises the community about 10 kilometers away from the main city streets. Arena Blanco is actually an island barangay at the mouth of Masinloc River, connected to Mampang by bridge. A sizable Badjao community resides here.

6. BFAR - Interviews with personnel from Bureau of Fisheries and Aquatic Resources (BFAR) IX were also done by the researcher.

C. Public Documents & Data

Information available through government agencies and several Non-Government Organizations are also utilized to reinforce gathered from the two methods mentioned above.

D. Target Respondents

In general, this study would target ordinary residents within the area of study, neutral of gender and within the range of ages where individuals are mostly expected to make a living, approximately from age 18 to 60.

VI. Findings and Conclusions

A. Focus Group Discussions Verbatim Report

1. Mariki FGD (See Appendix 1)

2. Recodo FGD (See Appendix 2)

3. Taluksangay FGD (See Appendix 3)

4. Basilan FGD (See Appendix 4)

B. Focus Group Discussions Syntheses

1. Mariki FGD (See Appendix 5)

2. Recodo FGD (See Appendix 6)

3. Taluksangay FGD (See Appendix 7)

4. Basilan FGD (See Appendix 8)

C. Focus Group Discussions Summaries

1. *Mariki FGD*

a) Fishing or activities related to the fishing industry was cited as livelihood of most residents in Mariki area.

b) Vendors and laborers are also prevalent.

c) Sikad-sikad and jeepney drivers are also mentioned.

d) Financial aid or grants such as micro-financing most specifically are requested as form of government assistance, to buy materials for fishing and for vending.

e) Specifically cited is for purchase of motorized Bangka and fishing implements.

f) Scholarships for the youth.

g) The need for further effort in peace and order by the government in the area, so that business activity would be unhampered.

h) Cited also fairness and equality in employment in government for Muslims.

i) There was emphasis on discrimination against Muslims in terms of job employment and in grant of privileges, like scholarships and livelihood programs.

j) Majority of respondents wants peace and unity.

k) One respondent stated that independence for Mindanao should be granted in order that peace will prevail in Mindanao.

l) Most of respondents are disinterested in the issue of peace process, especially on the topic of MOA-AD.

m) Most youths do not go to college for financial and economic reason.

n) College graduates in the area remains unemployed in their respective fields, like criminology as mentioned.

o) There are no cooperatives or community organizations known by the respondents.

p) Computer and Internet knowledge are mostly culled from Internet cafes by youths.

q) Television ownerships is very prevalent.

r) Private ownership of vehicles is minimal.

s) Land ownership is minimal.

2. *Recodo FGD*

a) Fisihing or fishing-related industry is cited as mostly the livelihood of residents in Recodo.

b) There are a number of the populations working in sardine factories that are situated within the vicinity.

c) Women usually go to vending and buy and sell activities.

d) Majority of respondents complained about the presence of foreign fishing boats as they compete directly with small fishers in the area for catch.

e) Most participants stated 'capital' or financial grant as most needed assistance from the government, mostly for purchase of fishing materials, like motorized banca and for start-up good in vending.

f) Scholarships for the youth is also mentioned with statement that present scholarship programs of the government for Muslim minorities are insufficient and ineffective as targeted beneficiaries are not served mostly.

g) Livelihood trainings and seminars are also cited as necessary livelihood assistance.

h) Most respondents want peace agreement to be resolved, thru dialogue.

i) Some stated dissatisfaction with the ongoing peace process citing how it did not improved the lives of Muslims despite years of negotiations and implementations.

j) One respondent stated that independence should be granted Mindanao, citing that what belongs to the Mindanaoans should be given to them.

k) Others want to have 'peace and unity'.

l) There was emphasis on discrimination against Muslims in terms of job employment and in grant of privileges, like scholarships and livelihood programs.

m) Most youths do not go to college due to economic restraints.

n) Most kids go to school upto highschool level.

o) Computers are taught in elementary school level.

p) Youth gain knowledge of computer use and Internet thru Internet café.

q) Land ownership is minimal.

r) Residential lot distributed thru urban poor mortgage program.

s) Cooperatives and community organizations are inactive due to general disinterest of members and lack of support from authorities.

t) Local Government Unit is supportive in terms of infrastructure projects, community participation and in governance.

3. *Taluksangay FGD*

a) Participants agree on the fact that mostly in Taluksangay, residents are depending on seaweeds farming and fishing.

b) The need for alternative livelihood as seaweeds farming is often seasonal in harvest.

c) There is a lengthy period of waiting time before agar-agar is mature enough for harvesting.

d) Financial aid or grant as general idea of effective assistance from the government.

e) Lack of management or technical support cited for failure of livelihood programs before in their area.

f) In the nearby Sacol Island, there is a sizable number of farmers, mostly coconut farmers while others are also in seaweeds farming.

g) Most participants agree that financial assistance to seaweeds farmers and fishermen are the best means of help from the government, allowing them to buy more materials and seedlings for increasing seaweeds production.

h) In the case of fishermen, cash assistance will aid them in purchasing more effective implement like a motorized Bangka and fishing nets.

i) Participants state that there was no such assistance from the government before.

j) There were also no cooperatives established in the area.

k) Income depends mostly on size of plantation area.

l) Education remains a serious problem in the area.

m) Youth mostly stopped at high school for lack of financial resources.

n) Scholarships are not sufficient since the main concern is the daily expenses of schooling like fare and meal allowances.

o) General sentiments towards the need for peace

p) Lack of livelihood is pointed out as the main cause for lack of peace.

q) Participants agree that peace dialogue is the best means for resolving the conflicts in Mindanao.

r) Most youth do not go to college due to financial concerns.

s) Others who were successful had transferred residences.

t) Computer knowledge is taught in highschool.

u) Youths learned to use computer and Internet through cafes.

v) Most residences own television units.

w) Transportation is mostly thru public transport.

x) Land ownership is not applicable since most houses are situated in 'free area'.

4. *Basilan FGD*

   a) Fishing or fishing-related industry is cited as mostly the livelihood of residents in Basilan.

   b) Seaweeds farming become an alternative livelihood for farmers waiting for harvest time.

   c) A sizable number of the Muslim populations are involved in farming activities mostly on abaca, copra and cassava.

   d) Education was cited as primary concern in Basilan as schools are not accessible in far-flung areas.

   e) Financial grant was mentioned for livelihood assistance.

   f) Lack of hospitals was also cited.

   g) Scholarships for the youth is also mentioned with statement that present scholarship programs of the government for Muslim minorities are insufficient and ineffective as targeted beneficiaries are not served mostly.

   h) The need for technical and management assistance in terms of projects, such as introduction of new technologies in aquaculture.

   i) Education is at a minimal as merely 30% goes to complete college education.

   j) Computer and Internet knowledge is at a minimal.

   k) Land ownership is high.

   l) Private vehicle ownership very mow.

   m) Television as a form of media is very low due to lack of electricity.

   n) OFW's comings from the area are mostly domestic helpers.

   o) Most respondents categorically desire peace when asked about the issue of peace process in Mindanao.

p) For peace process to succeed all concerned parties should be involved, like the MNLF and Muslim traditional leaders.

q) Others want to have 'peace and unity'.

r) Independence for Mindanao as not tenable for being not ready and transitional phase was mentioned as a solution.

D. Interview Verbatim Report

1. Siromon Islands (See Appendix 9)

2. Campo Muslim (See Appendix 10)

3. BFAR Personnel (See Appendix 11)

4. Arena Blanco (See Appendix 12)

E. Interview Summaries

1. Siromon Islands Interview

a.   Most of the residents in Siromon Islands and the other parts of 11 Islands are fisherfolks with minority comprising farmers.

b.   Fishing methods are mostly thru primeval method like hook-and-line, fish-trapping and thru fishnet. Fish-caging are done sparsely and crab culture is still on the very initial phase.

c.   Seaweeds farming are not so prevalent due to open sea location.

d.	Farming in nearby Sacol Island is more active. Crops involved are coconut, cassava, corn and mango – among others.

e.	Most respondents opt for financial assistance as mode of help from the government, for the purpose of purchasing materials for fishing like boats and nets.

f.	The call for assistance in water supply and electricity is prevalent among respondents.

g.	The need for motorized banca was emphasized by most fisherfolks interviewed.

h.	Awareness of the Mindanao peace process among the respondents is low.

i.	Peace and unity were mostly stated by respondents when asked about the peace process.

j.	Access to education in the area is very minimal as the nearest elementary school is still one boat ride away.

k.	Water and electricity is major problem in the area.

2. *Campo Islam Interviews*

a.	Most of the residents in Siromon Islands and the other parts of 11 Islands are fisherfolks with minority comprising farmers.

b. Fishing methods are mostly thru primeval method like hook-and-line, fish-trapping and thru fishnet. Fishcaging are done sparsely and crab culture is still on the very initial phase.

c. Seaweeds farming are not so prevalent due to open sea location.

d. Farming in nearby Sacol Island is more active. Crops involved are coconut, cassava, corn and mango – among others.

e. Most respondents opt for financial assistance as mode of help from the government, for the purpose of purchasing materials for fishing like boats and nets.

f. The call for assistance in water supply and electricity is prevalent among respondents.

g. The need for motorized banca was emphasized by most fisherfolks interviewed.

h. Awareness of the Mindanao peace process among the respondents is low.

i. Peace and unity were mostly stated by respondents when asked about the peace process.

j.   Access to education in the area is very minimal as the nearest elementary school is still one boat ride away.

k.   Water and electricity is major problem in the area.

3. *Arena Blanco Interviews*

   a.   Most respondents are involved in fishing and seaweeds farming.

   b.   Most respondents request for financial assistance as aid from government.

   c.   Most youths there have no access to college education.

   d.   A patent lack of interest in the ongoing peace process.

   e.   Lack of water is a major problem there.

   f.   Most individuals go directly into fishing than seek education.

4. BFAR Interviews

   a.   BFAR personnel conduct livelihood and technical training to fisherfolks and seaweeds farmers.

   b.   Devolution of BFAR powers and prerogatives to LGUs in 1988 has rendered many programs ineffectual.

   c.   Many LGUs gives low priority to fishing industry.

   d.   Aquaculture should be encouraged nowadays due to overfishing in the seas.

   e.   BFAR introduces new aquaculture technologies to fisherfolks such as Tilapia raising and blue crabs culture.

   f.   BFAR distributes for free fishing materials and seeds for fisherfolks and seaweeds farmers.

   g.   Lack of funding limits BFAR capabilities to help and assist fisherfolks.

F. Overall Summary

Based upon the data and information referred to above, as gathered through Focus Group Discussions in selected Muslim communities in Zamboanga City and Basilan, as well as through one-on-one and group interviews, the following general findings were gathered.

1. Most Muslims in Zamboanga City and Basilan are involved in the fishing industry and seaweeds farming on an equal ratio. About 20% are into farming.
   The methods and means of their trade and occupation are still based on very old and very basic methods such as by hook-and-line and fish-trapping using make-shift cages. Seaweeds farming has very high volatility rate for diseases and rejections due to improper harvesting and inputs of weight-enhancing additions such as salt and cement.

2. Most fisherfolks and seaweeds farmers have opted for direct financial aid from the government as the most effectual mode of assistance from the government. The financial resources requested are mostly aimed at buying materials for their livelihood activities such as motorized banca for fisherfolks and seedlings for seaweeds farmers. Request for scholarships were also emphasized during the discussions and interviews as well as for basic necessities such as medical aid, water supply and electricity.

3. Most fisherfolks and seaweeds farmers who have been benefitted from government technical assistance have made statement about the lack of management and marketing component of the programs. They suggested that future training programs should have focused sustainability and management component.

4. Most Muslims in Zamboanga City and Basilan prefers dialogue and peaceful resolution to the ongoing peace process in Mindanao with a sizable portion of respondents showing lack of interest in the issue. It is observed that primary of their concerns are their livelihood and conducive environment for everyday living.

5. Most respondents and participants in this study --- Muslims in Zamboanga City and Basilan --- are experiencing high incidence of deficiency in wealth and resources, living very simple lifestyles and mostly in poverty and depending on non-permanent incomes, without fixed daily profit, resulting to very low educational level or attainment for most of them.

6. Lack of access to education is a general and very prevalent condition in all Muslim communities in the study area mostly due to financial constraints. Scholarship programs currently in place are not making inroads on this problem and in fact are so insufficient to allow youths to enter college.

G. Conclusion and Recommendations

1. Primary of recommendation is state-intervention in the primary livelihood activities of respondents and participants, including primarily technical and management assistance. Most emphasized complaints include government technical assistance that are short-lived and lack follow-throughs. A more focused and widespread intervention is necessary, updating capabilities of fisherfolks and seaweeds farmers, aiming at increasing productivity and thereon their income-generation capacities.

2. Modernization and introduction of latest aquaculture technologies should be a viable government intervention, introducing updated means and method as well as tools and instruments. Overfishing in the seas have rendered fish supply to sizable reduction in recent years, with the presence of modern commercial fishing boats which directly affects fish catches of small-scale fishing sector.

3. A widespread modernization and enhancement program for the fishing and aquatic industry by the government should be most timely at this time as this would inure

directly towards the betterment and upliftment of the lives and experiences of Muslims in Zamboanga City and Basilan.

4. There should be emphasis on the management component of every technical and livelihood assistance given by the government to fisherfolks and seaweeds farmers in order to have high-sustainability and success rate.

5. Values and behavioral enrichment among fisherfolks and seaweeds farmers are also highly recommended as it was observed that lack of productivity could be pointed out to work attitude and discipline in trade, where for example, some seaweeds farmers have lost their reputation for having been caught to have inputted alien materials in their produce, improperly enhancing weight. The researcher also observed that most fisherfolks are merely contented on their catches, without aim at increasing productivity or finding time to find means of improving their profit, by seeking attention of government agencies such as BFAR or the Department of Agriculture.

6. A more comprehensive scholarship program for Muslims and indigenous people should be initiated in place of the current program. Current scholarship programs are insufficient and inadequate, aside from not being responsive to the targeted beneficiaries.

VII.   Bibliography

1. Lloyd G. Reynolds, "ECONOMICS : A General Introduction", Yale University, HOMEWOOD Publishing, Illinois 60430

2. David Ricardo, "On the Principles of Political Economy and Taxation", London: John Murray, 1821. First published 1819.

3. Bala Balassa and Associates, Studies in Trade Liberalization (Baltimore, John Hopkins Press, 1967)

4. Porter, M. E. *The Competitive Advantage of Nations*. New York: Free Press, 1990. (Republished with a new introduction, 1998.)

5. R.C. Mascarenhas, "Comparative Political Economy of East & South Asian : A Critique of Development Policy and Management", Great Britain, Macmillan Press LTD 1999

6. Adam Szirmai Economic and Social Development : Trends, Problems, Policies, Prentice Hall, Europe 1997

7. Harry J. Waters CHINA'S ECONOMIC DEVELOPMENT STRATEGIES FOR THE 21st CENTURY, Quorom Books, 88 Post Road West, Westport, CT 06881 1997

8. WIKIPEDIA : http://en.wikipedia.org/wiki/Comparative_advantage

9. WIKIPEDIA : http://en.wikipedia.org/wiki/Vernon_Wesley_Ruttan

10. City of Zamboanga LGU official website : URL http://www.zamboanga.gov.ph/

11. ACT FOR PEACE PROGRAMME URL

   :http://www.actforpeace.ph/component/option,com_frontpage/Itemid,1/